Vanessa Lessa Azevedo
Ricardo Dobrovolski

Mapping bat sensitivity to wind farms in Brazil

Vanessa Lessa Azevedo
Ricardo Dobrovolski

Mapping bat sensitivity to wind farms in Brazil

A deductive modelling framework based on species functional attributes and habitat characteristics

ScienciaScripts

Imprint

Any brand names and product names mentioned in this book are subject to trademark, brand or patent protection and are trademarks or registered trademarks of their respective holders. The use of brand names, product names, common names, trade names, product descriptions etc. even without a particular marking in this work is in no way to be construed to mean that such names may be regarded as unrestricted in respect of trademark and brand protection legislation and could thus be used by anyone.

Cover image: www.ingimage.com

This book is a translation from the original published under ISBN 978-613-9-74831-0.

Publisher:
Sciencia Scripts
is a trademark of
Dodo Books Indian Ocean Ltd. and OmniScriptum S.R.L publishing group

120 High Road, East Finchley, London, N2 9ED, United Kingdom
Str. Armeneasca 28/1, office 1, Chisinau MD-2012, Republic of Moldova, Europe
Printed at: see last page
ISBN: 978-620-6-33388-3

ACKNOWLEDGEMENTS

I would like to thank the CAPES funding agency for providing financial resources for the development of the project.

To the Postgraduate Programme in Ecology and Biomonitoring of UFBA, to the Institute of Biology and to all the professors.

To my supervisor, for helping me with my project and to my colleagues in the Ecology and Conservation Laboratory, especially Daniel who always helped me in moments of despair, Myrla and Chris.

I thank the members of the panel for accepting our invitation and thus contributing to the improvement of this work.

To my classmates who were very important in this process. To Ricardo Negrão who helped me in the use of ArcGIS.

To Daniel for putting up with all my moments and supporting me throughout these two years and to my family, my mum and my siblings for all their love and support.

And I also thank my dear father who, although he is no longer with us, has always been my greatest example and supporter in relation to studies.

TABLE OF CONTENTS:

SUMMARY

The high mortality of bats due to impacts on wind farms has raised concern worldwide. This phenomenon may be even more serious in a megadiverse country with important knowledge gaps like Brazil. In order to create a tool to help reduce this conflict, we generated a map of the sensitivity of bats to the impacts of wind farms and evaluated the areas of conflict in the current and expansion scenarios of this sector. For this, we performed a deductive modelling of habitat suitability that allowed us to refine the information on the distribution of these animals in the country, and we analysed functional attributes and landscape characteristics that are related to the risk of impact of these developments. In addition, we overlaid the sensitivity mapping with the location of wind farms present in the country and with the areas of wind potential. Our results showed that the Caatinga region deserves special attention in the current scenario, considering that among the areas evaluated, this biome was the one that presented the greatest conflict in the current scenario related to the locations with wind farms. Regarding the risk assessment in the expansion scenario, we observed, through a spatial correlation between the sensitivity mapping and the areas of wind potential in the country, we found a negative correlation, demonstrating that there is no conservation conflict, since the areas that presented greater sensitivity, are not the places that have greater wind potential, such as the North region that proved to be the most sensitive, but does not have high potential for wind production. Among the areas of conflict, we found that 0.5% of the areas evaluated represent sites with high sensitivity and high wind potential. Based on this result, we suggest that it is possible to expand the sector without generating great risk for the country's bats if there is adequate planning for this expansion. In addition, our study indicates areas that need more attention, and consequent rigour in the licensing criteria for future developments.

Key words: conservation conflicts; energy matrix expansion; bat impact sensitivity mapping; deductive habitat modelling;

CHAPTER 1

1. INTRODUCTION

Human population growth and rapid economic development have led to increased energy demands around the world. As a result, forecasts for the year 2040 have shown the possibility of an increase of up to 80% in energy demand and the continued dominance of fossil sources (IEA, 2014). This reality, coupled with current data on pollution, climate change and declining oil sources, has highlighted the need for new renewable energy sources (IEA, 2014; WWF, 2015).

In this context, wind energy represents one of the main sources of renewable energy in the world, due to its low emission of carbon dioxide (Jacobson, 2009). In Brazil, technological advances and government incentives have provided more and more resources to this sector, generating a wide expansion in the country (ABEE, 2013), and may triple production from 2014 to 2019 (Bernard et al., 2014). This type of wind power development has also benefited from the wide availability of areas with wind speeds suitable for the implementation of this type of energy source. Studies by Atlas Potencial Eólico do Brasil indicate more than 71,000 km^2 with suitability for this type of enterprise (Amarante et al., 2001).

Despite being considered a renewable energy source, this type of development is not harmless to biodiversity. Wind farms have impacts on wildlife, especially on populations of winged fauna (Sovernigo, 2009; AWWI, 2015; Arnett et al., 2016). At first, only the consequences generated to birds were known, however it is already demonstrated that the impacts caused to bats can be greater, demonstrating high annual mortality rates in different locations around the world (Kunz et al., 2007; Baerwald and Barclay, 2009; Arnett and Baerwald, 2013; AWWI, 2015; Arnett et. al., 2016). This reality generates an ecological, social and economic concern considering that the exact magnitude of this impact is still unknown, which generates a threat to the ecological services provided by these animals (Quinn et al., 2011).

Impacts to bats are not limited to direct effects generated by collisions or barotrauma with wind tower blade structures. And these impacts to bats do not occur randomly, as mortality samples do not directly reflect local faunal composition (Jonhson et al., 2004; Kunz et al., 2007; Arnett et al., 2016).

In addition, many species are affected indirectly through habitat loss and fragmentation of areas altered for access roads and wind tower construction (AWWI, 2015; Arnett et al., 2016). These impacts affect species in different ways and intensities depending on life history traits or functional attributes such as foraging types, flight types, mobility and habitat specialisation (Hull and Cawthen, 2013; Rydell et al., 2010; Barros et al., 2015; Rodríguez-Durán and Feliciano-Robles, 2015; Arnett et al., 2016). Therefore, the combination of species distribution data and their functional attributes can help to assess sensitivity to impacts generated by the implementation of wind farms (Quinn et al.,

2011; McGuinness et al., 2015). Some studies have shown that using these attributes together with landscape features can promote impact risk management (Okey and Kuzemchak, 2009; Quinn et al., 2011).

Bats make up the second richest group of mammals, constituting about 25% of this class with more than 1150 species recorded worldwide (Simmons, 2005). In Brazil this group is distributed in nine families, represented by 180 species (Nogueira et al., 2014; Feijó et al., 2015; Moratelli and Dias, 2015). This variety of species is accompanied by a wide diversity and ecological importance, which makes them key elements within ecosystems where they interact with a wide variety of organisms (Nogueira et al., 2014; Voigt and Kingston, 2015). Among their importance are pollination, seed dispersal, forest regeneration, pest control (Peracchi et al., 2007), in addition to being prey and predators of other animals (Kalka et al., 2008) and having an important epidemiological role since they can have a haematophagous habit (Bernard, 2002).

Despite their great importance, studies of bats in Brazil have only become more common in the last fifteen years (Pacheco et al., 2008). However, the data base regarding the composition and ecology of these species is still unsatisfactory (Bernard et al., 2011). As a consequence, many areas of high wind potential in Brazil represent sites with no data on bat species composition (Bernard et al., 2014).

In order to contribute to the reconciliation between the expansion of the wind energy sector and the conservation of winged fauna, some strategies have been demonstrated as important tools in reducing this conflict (Arnett et al., 2016; Guinness et al., 2015), such as impact sensitivity mapping based on key criteria related to the impacted fauna (Okey and Kuzemchak, 2009; Quinn et al., 2011); Santos et al.,2013; Rosconi et al., 2013; Roscioni et al., 2014; McGuinness et al., 2015) and with the landscape in question (Quinn et al., 2011; Santos et al.,2013; Rosconi et al., 2013; Roscioni et al., 2014).

The expansion of the wind energy sector in a way that does not generate deleterious effects on local diversity requires efficient management of the impacts generated by this sector (AWWI, 2015; Arnett et al., 2016). In this context, it is essential to assess whether geographical patterns of human development coincide with areas of importance for biodiversity conservation, since when both interests are at odds, conservation conflicts arise (Redpath et al., 2013). As a result, scientists have a key role to play in understanding the root of such conflicts in an attempt to generate techniques that test mitigation alternatives to help balance the two objectives (Redpath et al., 2013).

Our aim is to map the sensitivity to the installation of wind farms along the Brazilian territory, considering the distribution of bats, their functional attributes and landscape characteristics. In addition, we will test whether there is a conservation conflict between areas with high wind potential and areas with high level of sensitivity. Thus, this study aims to contribute to the planning of the construction of new enterprises and to the screening of areas that deserve greater attention in the evaluation of possible environmental impacts, serving as a tool for environmental agencies and decision makers.

CHAPTER 2

2. MATERIAL AND METHOD

2.1   STUDY AREA

The assessment of the potential susceptibility of bats to wind power developments was carried out throughout Brazil, representing an area of 8,515,767.049 km^2 (Diário Oficial da União, 2016) (33°45' to 5°16'N; - 34°47' to 73°59'W), where much of the national territory has wind power potential which favours the implementation of this type of development in almost the entire country (Amarante et al., 2001). On the other hand, the country has a high diversity of bat species (Nogueira et al., 2014) and with almost no information regarding impacts related to Neotropical areas (Sovernigo, 2009; Arnett et al., 2016; Barros et al., 2015; Rodríguez-Durán and Feliciano-Robles, 2015; Pacheco et al., 2014). For this reason, studies related to the interaction of bats with this type of enterprise are among the ten most relevant issues for the conservation of this group in the country (Bernard et al., 2012). Demonstrating the need for strategies that try to reconcile the expansion of this sector with the conservation of biodiversity in this country.

2.2   DATA

2.2.1 Species distribution

Of the 180 bat species recorded in Brazil (Nogueira et al., 2014; Feijó et al., 2015; Moratelli & Dias 2015) there are distribution data for 157 species, all recorded by the International Union for the Conservation of Nature - IUCN (http://www.iucnredlist.org). The 23 missing species were recently described and have not yet been assessed or have not yet had their geographical distribution recorded in IUCN. Consequently, for these species different methodologies were used to create distribution ranges, varying according to the amount of occurrence records available in the literature. For the eight (8) species that had more than five (5) records, the minimum convex polygon method was used, the same method used by the IUCN (NSW Scientific Committee, 2012). For the 3 (three) species with amounts of records that varied from 4 to 3 record localities, polygons were created and for the 12 (twelve) species with only 1 or more nearby records, a buffer was made around the centroid of the record points, this procedure was based on the delimitation of the smallest distribution area of the species provided by the IUCN, generating delimitations based on this parameter, all these procedures were performed in R, version 3.1.2 (http://www.R-project.org).

2.2.2 Data refinement (Deductive modelling)

To be able to create an efficient conservation tool, it is necessary to have at least knowledge related to the species composition of the evaluated areas (Rondinini et al., 2011). To assess the potential susceptibility of bats to wind energy developments in Brazil, it was necessary to assess the geographical occurrence of bat species present in the country. In view of the lack of information regarding species composition in the country, which is still unsatisfactory (Bernard et al., 2011), the variable quality and the lack of security derived from the incomplete evaluation of the records made locally (Boitani et al., 2011), it is necessary to seek strategies that overcome problems arising from errors of commission or overprediction within the areas evaluated, generated by the Wallacean deficit, which is related to the knowledge gaps regarding the distributions of the species (Lemes et al., 2011).

In this sense, in order to override the problems arising from the Wallacean deficit, we used deductive suitability modelling, which allows the use of databases available at coarser resolutions such as extent of occurrence, accessed by the global mammal data available from the IUCN Red List of Threatened Species (www.iucnredlist.org), together with habitat preference information (IUCN) which was supplemented with information available in the literature. By utilising these data with information regarding landscape features we identified unsuitable habitats within the species' range, thus improving data resolution (Rondinini et al., 2011). To this end, we developed a suitability model in grids with a resolution of 0.1° x 0.1° for each species.

To model the distribution area of the Brazilian bat species assessed by IUCN, the species distribution data were obtained in vector format (ESRI shapefile), which was used to obtain the presence matrix. This matrix was used together with the habitat preference matrix, where we used as a basis to reclassify the land use map that was obtained from the *Global Land Cover Map* 2009, a global mapping performed through fine-scale satellite imagery (ESA, 2010). This map was then reclassified from the original land cover classes of the *UN Land Cover Classification System - LCCS* (Di Gregorio and Jansen, 2000) based on the categories of the habitat preference matrix. Thus, the distribution of species was refined, subsequently generating the thematic bat richness map with the geographical distribution of all species provided by the IUCN with grids at 0.1° x 0.1° resolution.

2.2.3 Matrix of functional attributes of species

Measurements of functional attributes can help predict the effects of human alterations (Cianciaruso et al., 2009). And the combination of these functional attributes with species distribution data can assist in assessing sensitivity to impacts generated by the implementation of wind farms (Wilman et al., 2014; ICMBio, 2016). In this context, we assessed the relative weight of a trait in relation to its ability to predict mortality risk. To this end, we used morphological and ecological characteristics of the species that have been identified in the literature as responsible for increasing their predisposition to local extinction risk (Jones et al., 2003; Safi & Kerth, 2004; Sagot & Chaverri, 2015, Farneda et al., 2015) or to risk of impact to a certain type of structure such as towers or wind blades (Bradbury et al., 2014; Ferreira et al., 2015; Arnet et al., 2016).

The data referring to the key characteristics of the species will compose a matrix "species x attributes". In which for each attribute it must be weighted with the degree of susceptibility that must be established in order to measure the level of risk that a specific attribute generates in relation to the probability of impact resulting from the influence of that species characteristic. The values of susceptibility grades ranged from 1 for categories considered as low risk, 2 for intermediate and 3 for high.

A. Migratory habit

This trait has been shown in temperate zones to be a good predictor of impact risk in wind farms, migratory species have been shown to be the most impacted by this type of development (Bernard et al., 2014; AWWI, 2015; Arnett et al., 2016), making this group the most susceptible. Certain species that occur in Brazil are long-distance migrants, or are considered seasonal migrants in other countries (Milner et al., 1990; Arnett et al., 2016; Schuster et al., 2015). Coincidentally, these species are also recorded in collisions with wind blades in Neotropical regions (Barros et al., 2015; Rodríguez-Durán & Feliciano-Robles, 2015; Pacheco et al., 2014). For this reason, we consider "Migratory Habitat" an important measurement attribute for vulnerability of this type of species. Species with a record of migration in the literature received a weight of 3, as it represents a high-risk attribute, and non-migratory species received a weight of 0.

B. Food guild / foraging strategy

Bats exploit a wide variety of foods, and vary in where they forage and how they collect their prey. For this reason, food type and feeding mode have been used as important categories in structuring bats (Bernard, 2002; Safi & Kerth, 2004; Klingbeil & Willig, 2009; Farneda et al., 2015). These animals when classified as guild types can be: Open area aerial insectivores; Edge area aerial insectivores; Forest aerial insectivores (highly obstructed environment); Collecting insectivores (highly obstructed environment); Collecting piscivores (obstructed environment); Haematophages (obstructed environment); Nectarivores (obstructed environment); Frugivores (obstructed environment); Omnivores (obstructed environment) and Carnivores (obstructed environment) (Bernard, 2002; Kalko et al., 1996). These guilds tend to exploit different types of environments due to their evolution of flight and echolocation in which species that collect their food in the air use open environments or vegetation edges, while species that collect on the substrate tend to use more forested areas (Denzinger & Schnitzler, 2013).

Open-area insectivorous bats are the most directly affected by wind developments, with the majority of species recorded in collisions or barotrauma (Arnett et al., 2016; Schuster et al., 2015; Hull & Cawthen, 2013). These species received a weight of 3. Insectivorous and carnivorous species, known as collector animaivores, are considered more vulnerable to environmental changes due to their low dispersal capacity (Kalko et al., 1996; Safi & Kerth, 2004), and also received a weight of 3 as they are more vulnerable to indirect impacts generated by the implementation of new wind sector developments. Species representing the guild frugivores of highly obstructed environments received a weight of 2, as many of these species can travel long distances to forage (Kalko et al., 1996; Bernard, 2002), increasing the risk of being targeted by wind tower blades, even though they do not have other ecological characteristics described as vulnerable to such an impact (Barros et al., 2015; Rodríguez-Durán & Feliciano-Robles, 2015). The species of the guild edge aerial insectivore and piscivore gatherers were also weighted two because they may utilise more open vegetation environments such as vegetation clearings, which increases their risk of being impacted. The remaining guilds: Nectarivores; Omnivores; Haematophagous; Forest aerial insectivore, received weight 1.

C. Conservation status

Species were assessed according to the conservation categories described by the IUCN. Species in the Endangered (EN) and Vulnerable (VU) categories received weight 3, Data Deficient (DD) species, which can in many cases be considered as threatened or near threatened species due to the lack of information and the current risk of these species (Jetz and Freckleton, 2015), also received weight 3. Species in the Near Threatened (NT) and Least Concern (LC) category received weight 2 and 1, respectively. The Critically Endangered category was not represented among the species assessed.

D. Wing morphology

Wing morphology in bats is a good predictor of many ecological characteristics including habitat types, foraging strategies and *home range* size (Norberg and Rayner, 1987). In addition, wings are related to physiology, as wing shape is an adaptation that reflects the demand for flight performance when foraging under particular ecological conditions, and is related to the energy expenditure of movement (Norberg and Rayner, 1987; Norberg, 1994). Therefore, this attribute predicts different functional relationships, including the risk of extinction that may be related to habitat availability and/or accessibility (Safi and Kerth, 2004) and resilience to environmental changes (anthropisation), since certain aspects of the wing enable the displacement of long distances between shelter sites and foraging areas with low energy expenditure (Jung and Kalko, 2011). These attributes related to wing morphology may also favour the use of more open areas of the landscape (Neuweiler, 2000), and the increase in home range, which allows animals to move over larger areas between their shelter environment and the foraging areas used (Jung and Kalko, 2011), which makes some species more vulnerable to impacts with wind towers (Arnett et al., 2016).

Thus, these attributes allow us to recognise ecomorphological differences of bat species that reflect the increased vulnerability of species to the different types of impacts generated by wind farms, being for this reason a good predictor to assess the increased risk of impact (Arnett et al., 2016; Hull and Cawthen, 2013; Rydell et al., 2010). To be able to give different weights to the species related to these two attributes that are quantitative, a quartile distribution of the data was generated and the weights assigned ranged from 1, low risk, to 3, high risk (Tab.1).

D.1 Wing aspect *ratio* or wing profile (*Aspect ratio*)

The wing aspect ratio describes the shape of the wing: the wingspan squared divided by the wing area. This characteristic is related to flight aerodynamics and energy efficiency (Norberg and Rayner, 1987). Long and narrow wings are related to a high wing aspect ratio, which generates low energy expenditure. In contrast, bats with long wings tend to lose manoeuvrability leading them to use more open areas (Safi and Kerth, 2004; Bader et al., 2015), and higher strata of vegetation (Hull & Cawthen, 2013; Rydell et al., 2010), while species with low wing carrying capacity have high flight manoeuvrability, allowing them to use more obstructed environments such as the interior and edge of forests (Neuweiler, 2000). The upper quartile classes, which corresponded to 25% of the highest values of the distribution found in the species, received weight 3, due to the increased exposure in open areas, utilisation of higher altitudes (Norberg and Rayner, 1987). The lower class, which represented 25% of the lowest values, also received weight 3 due to increased sensitivity to habitat alteration (Safi and Kerth, 2004). The classes of values distributed above the 1st quartile and below the 2nd quartile were weighted 1, as these are species that tend to utilise more complex environments but not as restricted as the 1st quartile species, while the values distributed above the 2nd quartile and below the 3rd quartile were weighted 2 as they tend to utilise both types of environment.

D.2 *Wing loading* capacity

Wing loading capacity is related to the measurement of the wing surface in relation to the body weight: weight multiplied by gravity divided by the wing area. This attribute is positively correlated with minimum speed and negatively correlated with manoeuvrability (Norberg and Rayner, 1987). Species with high carrying capacity values have higher speed, providing high mobility and greater persistence in anthropised environments (Safi and Kerth, 2004; Bader et al., 2015), in contrast, these species are more exposed to impact with wind turbine blades (Arnet et al., 2016). On the other hand, species with low wing load capacity have lower movement speed, generating less flight autonomy and reduced dispersal capacity (Neuweiler, 2000), making them more vulnerable to habitat alteration (Meyer, et al., 2008). The upper class received weight 3, due to the use of open environments due to low manoeuvrability and the increase in living area (Norberg and Rayner 1987), which make them more vulnerable. The lower class, which represented 25% of the lowest values, also received weight 3 due to sensitivity to habitat alteration generated by the preference for more complex habitats (Safi and Kerth, 2004). The classes of values distributed above the 1st quartile and below the 2nd quartile were weighted 1, as these are species that tend to use more complex environments but not as restricted as the species in the 1st quartile, while the values distributed above the 2nd quartile and below the 3rd quartile were weighted 2 as they tend to use both types of environment.

E. Vertical stratification

The vegetation is composed of different vertical strata that have variations in physical and biological structures (Bernard, 2001), this reality provides an increase in local diversity enabling the coexistence of different species that have preferences for specific regions of the vegetation for foraging (Kalko and Handley, 2001). This type of attribute is strongly related to the risk of impact with wind blades, since species with a vertical pattern of habitat use of higher strata (emergent layers) use open area environments in different ways, either for displacement, foraging or migration, thus increasing susceptibility to collisions (National Research Council, 2007; Arnett et al, 2016; Hull & Cawthen, 2013) and for this reason received weight 3. Species that use intermediate strata, such as canopy and sub-canopy, may suffer collision with wind blades even in smaller proportions (Barros et al., 2015; Rodríguez-Durán and Feliciano-Robles, 2015), and for this reason they received weight 2. Forest species that use lower substrates, such as understory, are less subject to this type of direct impact, for example, since many species avoid non-forested areas and/or do not use the sweeping areas of the wind tower blade as much (Hull & Cawthen, 2013) and for this reason they received weight 1.

F. Shelter specialist

Studies have shown a positive relationship between habitat specialisation and extinction risk in several taxa (Safi & Kerth, 2004), including bats (Sagot & Chaverri, 2015). Species specialising in certain types of shelters have been shown to be particularly vulnerable to anthropogenic changes and human activities, in addition, shelters are extremely important for the survival and reproduction of these animals (Sagot & Chaverri, 2015), since these structures serve as a refuge against predators and climatic weather (Ferrara & Leberg, 2005), and are also extremely important for social relationships, such as copulation, feeding and maternity (Kerth et al., 2003). For these reasons, habitat loss and disturbances in colonies in shelters have currently been considered as one of the main causes of declines in bat populations (IUCN, 2016). In view of this reality, species considered as specialists received weight 3, species that used up to two sources of shelters received weight 2, and those that used more than three types received weight 1.

G. Cave

An important pattern observed in bat-related direct impacts from collision with wind turbine blades is the presence of species that utilise warm shelters such as caves (Rodríguez-Durán and Feliciano-Robles, 2015).

Despite the existence of few studies revealing collision patterns in Neotropical regions, this result may be ubiquitous in countries of this region including Brazil (Arnett et al., 2016). Species that use these environments such as representatives of the Mormoopidae and Phyllostomidae families have large aggregations and can utilise wide areas between shelter and foraging areas (Rodríguez-Dúran,

2009).

Table 1. *Summary of bat attributes with their respective divisions.*

	Attributes	Category	Weight
A	Migratory habit	Migration	3
		Resident	0
B	Food guild / trophic level	*IAA , IC*, C*	3
		F*; IAB*,PC*	2
		N; O; H*; IAF*	1
C	Conservation status	Least Concern (LC)	1
		Near Threatened (NT); Data Deficient (DD)	2
		Vulnerable (VU)	3
D.1	Wing morphology: Aspect ratio	Values between 1st and 2nd quartile	1
		Values between 2nd and 3rd quartile	2
		Values > 4th quartile; Values < 1st quartile	3
D.2	Wing morphology: Load capacity	Values between 1st and 2nd quartile	1
		Values between 2nd and 3rd quartile	2
		Values above the 4th quartile; Values < 1st quartile	3
E	Vertical stratification	Above the canopy	3
		Canopy	2
		Below the canopy	1
F	Shelter specialist	Expertise	3
		Qty of shelters up to 2	2
		Qty of shelters > 3	1
G	Cave	Yes	3
		No	0

* IAA- Open aerial insectivore; IC- Collecting insectivore, C- Carnivores; F- Frugivores; IAB- Edge aerial insectivore; PC- Collecting piscivores N- Nectarivores; O- Omnivores; H-Hematophages; IAF- Forest aerial insectivore.

2.2.4 Filling in missing data

The scarcity of data regarding the ecological characteristics of species has generated great difficulty in creating effective conservation strategies (Bernard et al, 2012). However, the exclusion of data due to lack of information has not been indicated, since it not only reduces the sample number, but also tends to generate a bias, which can lead to erroneous conclusions (Penone et al., 2014). While the use of missing data filling algorithms is a potential solution to this type of problem (Penone et al., 2014). In order to solve this information gap, the *missForest* function from the *missForest* package (Daniel and Stekhoven, 2013) in *R software* was used. This function is a non-parametric missing data filling algorithm ideal for cases of mixed data with categorical and continuous attributes such as the matrix in question. Only species characteristics with >50% of available data were used. To obtain the final weight regarding the sensitivity of each species to the risk of impact with wind farms generated by the functional attributes of the species present in Brazil, we created a matrix with a total of 2327 cells, of which 384 cells represented missing data. These data could not be obtained due to the scarcity of species-specific information (Bernard et al., 2012), especially for recently described or data deficient species.

To fill these missing data we used the *missForest* algorithm (Daniel and Stekhoven, 2013) to complete 164 data related to continuous values and 220 related to categorical data, which generated the complete data matrix. For this procedure we obtained an estimate of the normalised root mean square error (NRMSE) of 0.63 for continuous variables and an estimate of the proportion of falsely classified entries (PFC) of 0.15 for categorical data.

2.2.5 Environmental data

The likelihood of impacts derived from the implementation of wind farms is related not only to the characteristics of the species present at the site assessed, but also in relation to environmental characteristics (Arnett and Baerwald, 2013, Arnett et. al., 2016), such as land use, topography and vegetation cover (Arnett et. al., 2016). In addition, areas more suitable for foraging and/or shelter such as forested areas or more heterogeneous areas are considered more vulnerable to impacts than open areas (Rydell et al., 2010).

In order to make a sensitivity map that covers as broadly as possible the factors that influence the increase in risk of impact, environmental variables mentioned in the literature as important and that are available through quality data and with sample representativeness for the whole country were included in the evaluation in an equal way without generating a sampling bias. For this reason, we did not use maps of water bodies or cave areas, which despite being cited in the literature as influencing factors, there is no complete data for the entire national territory.

Certain characteristics of the environment have also been revealed as predictors in assessing the risk of impacts generated by habitat alteration (Voig & Kingston, 2015) or by the implementation of structures such as wind towers (Rydell et al., 2010; Piorkowski & O'Connell, 2010; Ferreira et al., 2015; Arnett & Baerwald, 2013). Given that bat mortality is not uniformly distributed, but rather spatially aggregated in areas with particular compositions (Quinns et al., 2011), environmental characteristics are capable of potentiating risks to bats (Piorkowski & O'Connell, 2010). The risk of bat collisions with wind energy production structures is also influenced by the interaction of bats with the landscape. Areas favourable for foraging or rich in shelters are considered to be more sensitive (Rydell et al., 2010; Piorkowski & O'Connell, 2010). For this reason, these sites are relevant for assessing the risk of impact of these animals with the environment (Ferreira et al., 2015). These attributes as well as the other spatial data were referenced by the WGS-84 coordinate system, as well as those related to environmental variables.

In order to generate weights of influence in the final map in an equal way among the variables, the score of the environmental variables was generated based on the final score generated for the functional attributes of the species. For this, a weight scale was used for the environmental factors based on the maximum value of the map generated for the sensitivity related to the functional attributes of the species. In this way, half of this score (680) was used to classify each environmental thematic map, thus generating a final map of sensitivity related to environmental factors with the same maximum weight as the functional attributes. For the weighting of the weights related to these factors, two thematic maps were used, one of land use and another of slope, both with grids with a resolution of 0.1° x 0.1°, described below:

A. Ground cover

We used a thematic map that was mapped using the Global Land Cover Map 2009 (ESA, 2010), a global map with 300m resolution. To aggregate values related to the land cover classes of the *Globcover* thematic map, values were related to the cover types regarding the influence of this environment in the potential increase of susceptibility to impact. The map was reclassified between high-risk and low-risk areas, where the maximum score values for habitat and minimum score values for areas not favourable for bats were measured (Tab.2.A). Classes of areas such as natural vegetation cover were considered high risk, since they are areas with greater availability of areas for foraging, shelter and desiccation (Arnett et al., 2016; Rydell et al., 2010). Classes such as urban areas, bare ground, anthropised environments, which are less suitable (Arnett and Baerwald, 2013; Piorkowski and O'Connell, 2010), were considered as low risk (Fig.1.A).

B. Slope

Regarding the topographic variable, the slope of the terrain was used. This environmental variable was used because it is considered important in the use of habitats by bats, since the slope of the terrain can influence the flight path. These can be directed towards landforms such as valleys and depressions (Jenkins et al., 2010), gaps in mountains, valley bottoms and ridges or mountain ranges (Bevanger, 1994). In addition to using geographical features of the landscape, such as hilltops as a reference for small passage routes between shelter and suitable foraging sites, it has been shown that species with functional traits targeted by impact have preferences for environments with greater slope than flat areas (Baerwald and Barclay, 2009). Thus, areas with greater slope are recognised to considerably increase the susceptibility to bat impact risk in wind farms, especially if they have nearby rock hills, as these sites are suitable for shelter formation for many species (Baerwald and Barclay, 2009; Arnett et al., 2016).

For the application of this criterion in Brazil, we used the Percentage Slope Map of the Brazilian Relief, made available by the Geological Service of Brazil - CPRM (http://www.cprm.gov.br), derived from the Digital Elevation Model (DEM), it offers a three-dimensional view of the area and optimises the verification of the variables that influence the performance of morphodynamic processes, prepared from the image obtained by the *Shuttle Radar Topography Mission - SRTM*. The variations of the slope values were classified according to the reference used by the Brazilian Institute of Geography and Statistics - IBGE (IBGE, 2007), in which values between 0 and 3% are considered Flat, from 3 to 8% Gentle wavy, 8 to 20% Wavy, 20 to 45% Strong wavy, 45 to 75% mountainous and > 75% Steep. In view of these classes, different weights were measured related to the probability of impact generated by the use of this environment, the mountainous and steep class received the highest weight, related to high risk, this score is related to the increase in areas with slopes and favourable areas for use, either by favouring the foraging of bats, or by the great availability of shelters (Arnett et al., 2016). The importance of these areas is also ensured by the classification of areas with these slopes as APP (Permanent Preservation Areas) according to the new forest code of Law No. 12,651/2012 (Brazil, 2012). The Strong Wavy class received half the maximum weight, which reflects intermediate risk, due to the presence of increased slope in these areas. Areas with lower slope, such as flat and gently undulating areas tend to have lower risk of impact (Baerwald and Barclay, 2009), for this reason they received minimum (Tab. 2. B; Fig. 1 .B).

Table 2: Summary of bat attributes with their respective divisions.

Environmental Attributes		Category	Weight
A	Ground cover	High risk - Natural Habitat	680
		Low Risk - No habitat	0
		Low Risk -Plan 0 to 3%	1
		Low Risk -Smooth undulating 3 to 8%	1

B	Slope	Low Risk -Wavy 8 to 20%	1
		Intermediate risk -Strong undulating 20 to 45%	340
		High risk - Mountainous 45 to 75%	680
		High risk - Steep > 75%	680

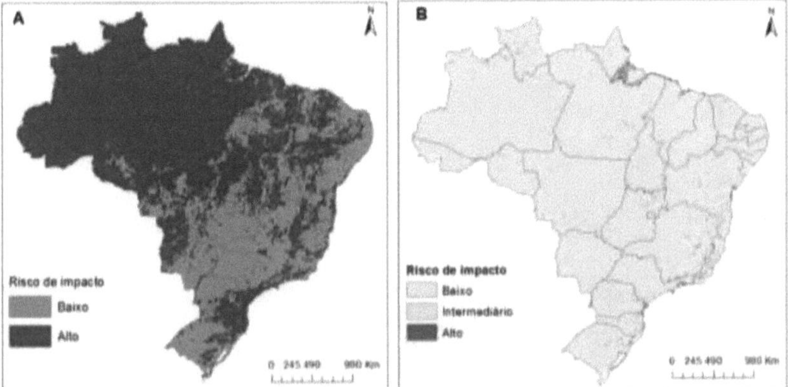

Figure 1 A- Impact risk map related to land cover for the territory of Brazil; **B** - Impact risk map related to terrain slope for the national territory. Data presented with cells with 0.1° latitude by 0.1° longitude.

2.3 SENSITIVITY MAPPING

By using the matrix related to the complete functional attributes it was possible to weight all the attributes to later add the sensitivity weight for each species, the total value of each species was then assigned to its respective distribution area. We then summed up all the distribution maps of the species present in the country generated by the respective overlap in the distribution of different species, which generated the sensitivity map related to the functional attributes of the species that increase the risk of impacts.

To make the sensitivity map related to the functional attributes of the species, we reclassified the geographical distribution areas of the 180 bat species evaluated, where the final geographical distribution raster of each species was represented by the final weight value of the respective species. This procedure was carried out by the spatial analysis tool through the "Reclassify" function of the *Raster* package (Robert and Hijmans, 2016) of R. Subsequently, these rasters were added through a map algebra, generating a final map with the sensitivity to impact related to the functional attributes of the species. To then be added to the sensitivity map related to the attributes of the species, in this process we generated the final map of sensitivity of bats in Brazil to the impact related to wind farms in the Brazilian territory.

To obtain the final sensitivity map, the maps resulting from the evaluation of the different predictor variables were used, being the same an overlap of the results found for the sensitivity based on the functional attributes of the species and the sensitivity generated for the two environmental variables. For this, a summation procedure was performed between the different maps, thus generating the final map of sensitivity of bat species in Brazil to impacts with wind farms (Fig.2).

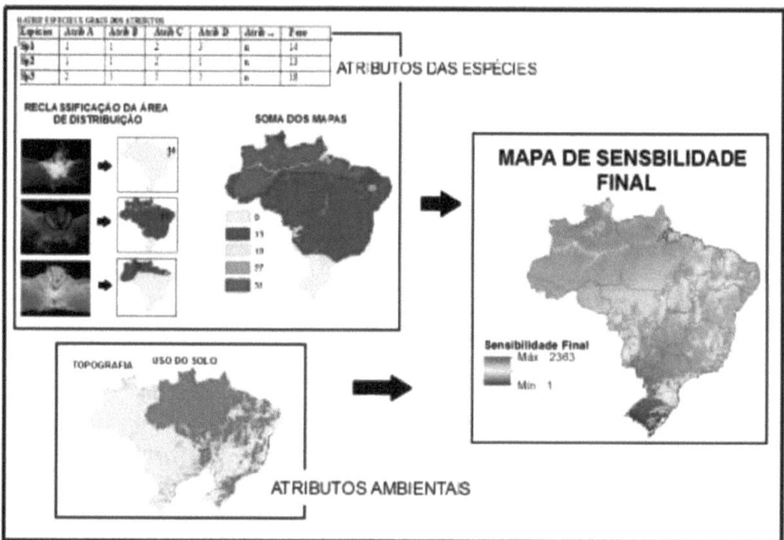

Figure 2 Schematic example of the method used to make the final map of sensitivity to wind impacts. In which the matrix of species attributes generated a final weight related to the risk of the species to the impact that was used for the reclassification of the species distribution range, these ranges were then superimposed, generating a sum of the weights in the locations from the different species present in each cell, as shown in the image of the sum of the maps. This map is then added to the environmental attributes to generate a final sensitivity map.

2.4 ANALYSES

Through the sensitivity maps related to the functional attributes of the species and the richness map, we performed a spatial correlation between the values of the cells of each grid of these maps in order to evaluate the spatial relationship between the variables. For this, the points were extracted, transforming the maps into an information matrix, which is composed of the set of coordinates of each centroid of the grid cells with the respective values assigned to each map analysed. These data had to be randomised and sampled with 5,000 grid cells to be analysed due to computational limitations. This analysis was performed in the SAM software ("*Spatial Analysis in Macroecology*", available at: www.ecoevol.ufg.br/ sam) (Rangel et al., 2010). This procedure was performed five times to ensure the result of the analysed samples, where the average of the five analyses was obtained as a result due to the similarity of the results. For this analysis we used the method of Clifford et al. (1989), which calculates the degrees of freedom corrected for the statistical test of correlation analyses, since geographic data are often spatially autocorrelated.

2.4.1   Areas of conflict

We performed an overlay between the final sensitivity map and the existing parks in Brazil in order to evaluate the areas with greater sensitivity and with the presence of wind farms that may indicate areas that need greater attention due to the risks generated to bat diversity.

In order to verify the potential risk areas related to the future expansion scenario of this sector, we performed a spatial correlation analysis between the data obtained from the final sensitivity and the existing wind potential in Brazil, which is based on the average annual wind speed for the Brazilian territory, from the Atlas of Wind Potential of Brazil (Amarante et al., 2001). To this end, these data were made compatible at a resolution of $0.1° \times 0.1°$ and placed in the same dimensions as the final sensitivity map to later extract the information matrix from them. The correlation analyses followed the same procedure as described above.

To quantify the areas of conflict (areas with high sensitivity and high wind potential), a scatter plot was made between the data where a threshold was measured based on the distribution of the data evaluated, where the distribution values present in the upper quartile were used because they contain the most critical values. To analyse the spatial distribution of these conflict areas, a bivariate map was made that simultaneously evaluates the two variables spatially distributed in quantiles, in order to reveal the relationship between them, this procedure was generated in the R *software*. The bivariate map shows the quartile of distribution that presented high values in both maps evaluated (Fig. 5 and 6), indicating their location on the map.

3. RESULTS

The recorded bat species richness (Fig. 4) as expected by the distribution pattern of other taxa demonstrated a wide heterogeneity in the country. Our result demonstrated a higher diversity in the northern region, with species richness decreasing towards the southern region of the country, exhibiting latitudinal gradient in the pattern of bat diversity in the country, with increasing species richness with decreasing latitude.

The number of species in a single cell ranged from 0 to 106, and most cells had a richness of 65 species.

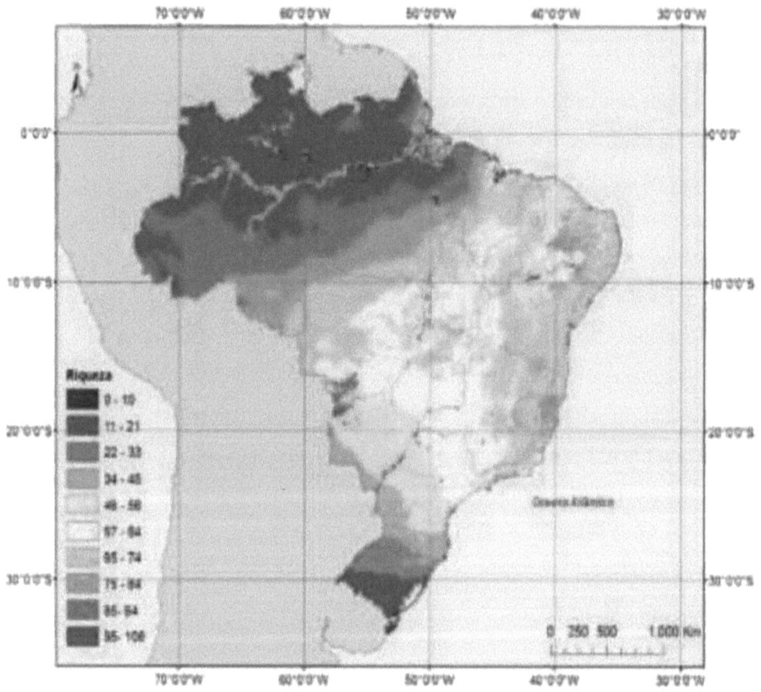

Figure 4: Map of bat species richness in Brazil. Data shown with cells at 0.1° latitude by 0.1° longitude.

The mapping of sensitivity related to the functional attributes of bat species (Fig. 5) also showed a latitudinal gradient, with higher sensitivity values in the north of the country. The distribution of the values of this sensitivity ranged from 0 to 1359, with a mean of 915 and a standard deviation of 245, the most frequent sensitivity value among the grid cells was 1125. This sensitivity when related to the distribution of species richness showed a positive correlation spatially (Pearson r

= 0.94, p<0.001).

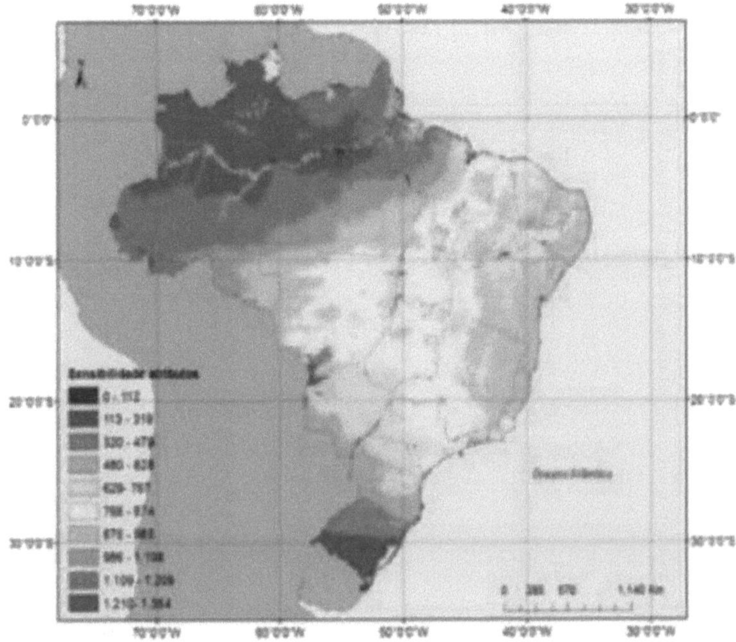

Figure 5: Sensitivity related to the functional attributes of bat species in the Brazilian territory that showed a relationship with the increased susceptibility to the impacts generated by the implementation of wind farms, whether direct or indirect. Data presented with cells with 0.1° latitude by 0.1° longitude.

The sensitivity map of bats to wind farms (Fig. 6) demonstrates the risk of impact to bat diversity generated by the presence of wind farms.

The sensitivity found among the grid cells ranged from 1 to 2363, the distribution of sensitivity presented a mean value of approximately 1400, while the most frequent sensitivity level was 1832, with approximately 25% of the cells with sensitivity higher than this value (Fig. 7).

The northern region showing the highest sensitivity in the country, the central-western region of the country showed higher levels of sensitivity in the northern part, while the northeast showed a wide range in the level of sensitivity with sites with higher levels in the central part, the southeastern region also showed wide variation in levels, with some points with high levels of sensitivity, such as Rio de Janeiro near the region of Petrópolis and Teresópolis, in the southeastern region of Espírito Santo that extends across the border in a part of the state of Minas Gerais, while the southern region presented the lowest sensitivity values, with only the state of Santa Catarina and southern Paraná with intermediate levels.

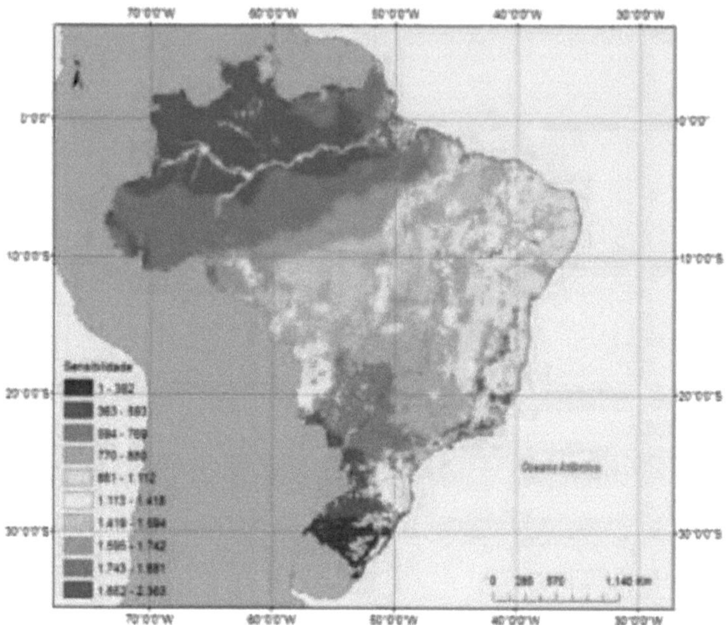

Figure 6: Sensitivity map of bats to wind farm developments taking into account the analysis of functional attributes of the species and assessment of risks generated by the landscape. Data presented with cells with 0.1° latitude by 0.1° longitude.

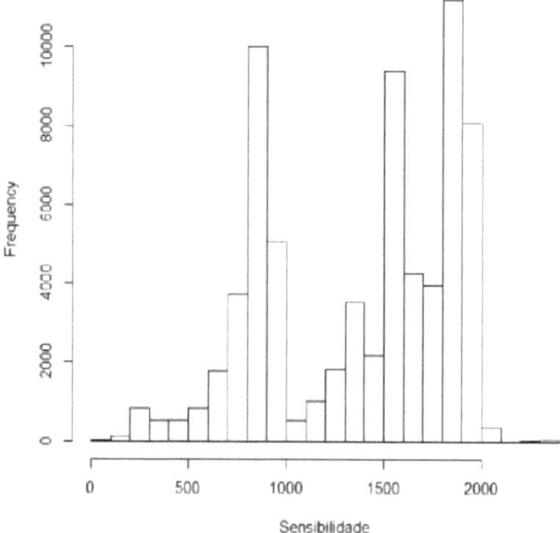

Figura 7: Distribution of data from the grid cells of bat sensitivity to wind farms.

According to the wind farms present in the current scenario (wind farms built or under construction) when superimposed with the sensitivity of bats (Fig. 8.A), the Northeast region presents the largest part of the farms at present with approximately 79% of the wind farms, in addition to presenting intermediate levels of sensitivity, with most of these developments in the region being inserted within the Caatinga biome (Fig. 8.B), representing the main area of conflict in the current scenario. current scenario. Among the states present in this area, Bahia presented the largest share of Brazil Parks, with approximately 28% of the parks, followed by Rio Grande do Norte with approximately 25% and Ceará with 15%.

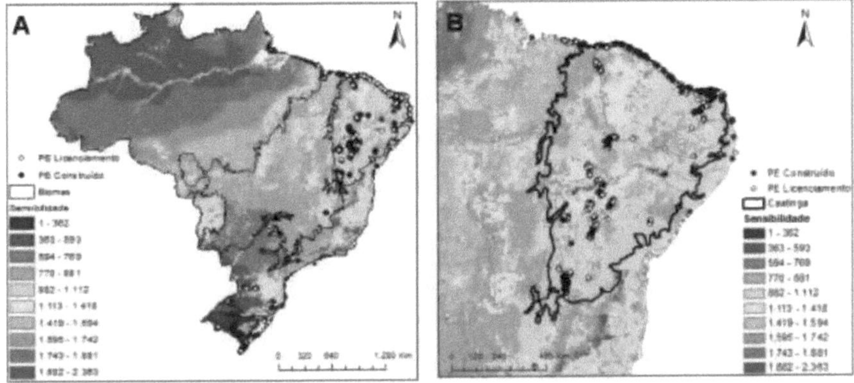

Figura 8: **A-** Map of bat sensitivity to wind farms with the overlap of Brazilian biomes and wind farms present in Brazil. Data presented with cells with 0.1° latitude by 0.1° longitude; **B** - Map of sensitivity of bats to wind farms for the Brazilian Northeast region with emphasis on the caatinga biome demonstrating the areas of overlap with wind farms. Data presented with cells with 0.1° latitude by 0.1° longitude.

We found a negative correlation between bat sensitivity and the future scenario of wind energy sector expansion (Pearson's r = -0.602, p < 0.001). However, 369 cells (0.5% of the total) showed simultaneously high sensitivity and high potential, representing conflict areas, being part of the upper quartile of the distribution of both variables simultaneously (Figs. 10 and 11).

Table 3: Average sensitivity of bats to wind farms and quantification of wind sector investment in different biomes of Brazil.

Biome	Average sensitivity	Built Parks	Licensed Parks	Wind energy investment (%)
Amazon	1721	2	0	0.15
Cerrado	1095	4	44	3.5
Caatinga	1142	227	732	70
Atlantic	1002	39	40	6

Forest				
Pampas	357	61	171	17
Pantanal	1178	0	0	0

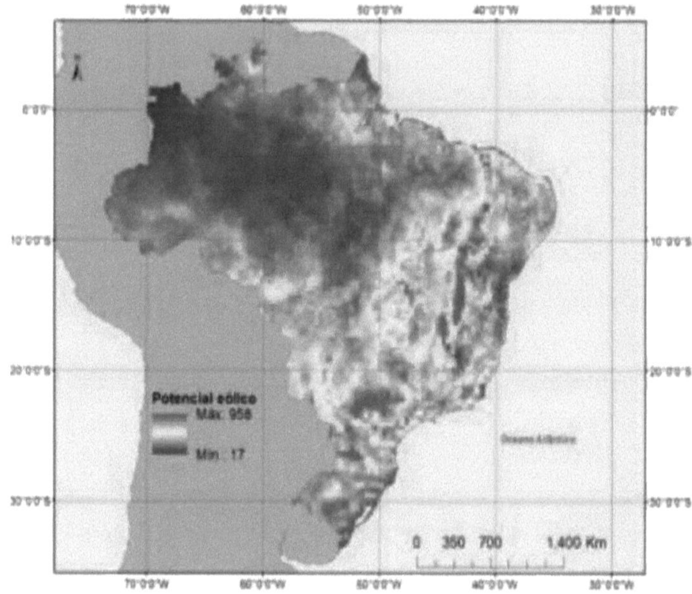

Figura 9: Estimated wind energy potential (in W/m2) in Brazil, based on data from the *Atlas of Wind Potential in Brazil* (Amarante et al., 2001). Data presented with cells at 0.1° latitude by 0.1° longitude.

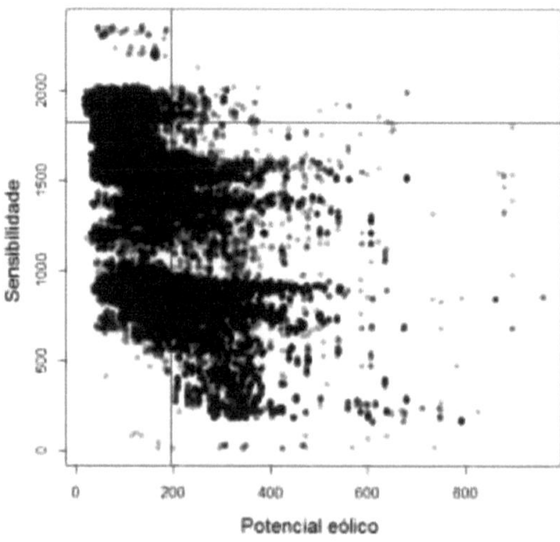

Figura 10: Scatter plot of sensitivity and wind potential data with threshold equivalent to the upper quartile

of the data shown by the vertical and horizontal lines respectively.

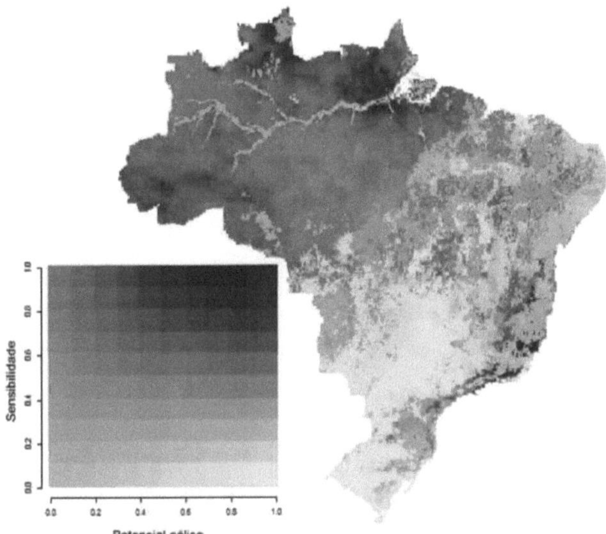

Figure 11: Areas of conflict relating bat conservation and wind production in Brazil. Each colour change means a 10% change in the quantile of our variables, highlighting in yellow the areas that presented high values of potential but low sensitivity, blue represents areas with high sensitivity but low potential, and in lilac conflict areas that presented high values for both variables, these regions represent the places that most deserve attention in the expansion of the wind sector in the country. Data presented cells with 0.1° latitude by 0.1° longitude.

4. DISCUSSION

The scenario of expansion of the wind energy sector in the country (EPE, 2015), associated with the scarcity of information related to the impact generated by wind farms on bats in a megadiverse place like Brazil (Bernard et al., 2014), makes studies that try to predict the risks generated by this enterprise to the biodiversity of bats fundamental for the conservation of this group in the country (Bernard et al., 2012). In this context, our study sought to create strategies to reduce environmental problems such as the risk generated by wind farms to bat conservation, in an innovative way. Although there is a set of knowledge gaps, based on the available data, we carried out an assessment of the risk of developments and conservation conflicts associated with this energy sector.

In this context, our study is a pioneer and differs from other existing maps of risk of sensitivity to impacts related to wind farms. Our mapping has a deductive approach (Mace et al., 2000), in which we made an assessment based on all bat species present in the country, taking into account the different types of impacts generated by this energy source and the characteristics of the landscape assessed. While other mappings used different approaches with inductive methodologies, such as the study carried out in Portugal, which performed a distribution modelling of four bat species recorded as impacted by wind turbines, data from collision records and landscape characteristics of the evaluated parks to then assess a probability of mortality and indicate the environmental factors that promoted the assessed risk (Santos et al., 2013). While another mapping carried out in Italy used bat species distribution modelling to overlap with the location of existing wind farms to access patterns in the alteration of favourable foraging habitats generated by present wind farms to identify areas vulnerable to the construction of future developments (Roscioni et al., 2013), this same method was used to identify areas vulnerable to the construction of future developments (Roscioni et al., 2013).), this same method was also used to evaluate the effect of wind farms on the connectivity of areas as a support to access risk areas, where the distribution of a migratory bat species was modelled to then create a connectivity map, in order to identify potential corridor areas for this species, which were later overlaid with existing wind farms and under licensing to verify the areas of greatest risk (Roscioni et al., 2014).

Other impact risk mapping studies have also been conducted with birds, some with approaches more similar to ours, such as the work of Quinn et al. (2011) that conducted a collision risk assessment with transmission lines in a local approach, focused on the group of freshwater waterbirds, which has greater vulnerability to impact, along with spatial characteristics of the landscape that were ranked based on the relationship with the increase in impact risk, generated by the assessment of collision areas recorded at the site. While the work of McGuinness et al. (2015), which also selected a set of specific species related to impact vulnerability. And used a method to rank species sensitivity similar to ours, but based on other species information. They also overlaid the weights of species present at the same site to generate a ranking of impact sensitivity without taking into account the characteristics of the landscape assessed. While another study modelled areas of conflict between wind farms and wildlife based on plant and animal species diversity layers available in Pennsylvania related to the conservation status of species present in the region. To verify the areas of conflict these data were overlaid with areas with high wind speeds (Okey and Kuzemchak, 2009).

Regarding our results, our richness data followed the typical global mammalian trend, showing higher diversity near the tropics, with a strong latitudinal gradient (Stevens, 2004; Buckley et al., 2010). We found a positive association between species richness and functional attributes. This result is probably related to the expected relationship between richness and functional diversity (Safi et al., 2011), considering that the functional attribute is an information used to measure functional diversity (Cianciaruso et al., 2009). This relationship shows that although functional attributes are important in this analysis, species diversity has great weight in impact sensitivity.

Our map of bat sensitivity to wind farms indicated areas in the northern part of the country and some sites in the south-eastern region were the most sensitive to wind developments. Despite the high levels of sensitivity indicated for the northern part of the country this region does not have high wind potential, with only two areas that stand out for having wind potential, the north of Roraima and Amapá. The Southeast also has a small region with wind potential. The rest of the country, with the exception of some other sparse locations in the national territory, presented lower sensitivity values.

The Northeast region, although not highlighted among the areas with greater sensitivity, deserves to be highlighted for the large investment in this sector, with the presence of many parks (Fig.8.B), among them the largest wind farms installed in the country (Bernard et al., 2014), this region represents one of the places with the greatest wind potential in the Brazilian territory. With a wind production forecast of up to 9 megawatts (MW) for 2024, it tends to be the area with the greatest expansion in the country (EPE, 2015). In this context, the Northeast proves to be the main area of conflict in the country today, with areas with intermediate sensitivity, presenting the highest rates among the locations with wind farm investments in the current scenario. In addition, most of the parks present in this region are in the Caatinga. This reality further emphasises the prominence of these areas, given that this biome is among the most degraded in the country, with more than 45% of its territory already deforested (Castelletti et al. 2004), only 1% of its territory is within integral protection units (Silva et al., 2004), in addition to the scarcity of information regarding the species present in these places (Bernard et al., 2011; Bernard et al., 2014), and the fact that it is an area with a large number of karst areas (Oliveira-Galvão, 2014), a fact that suggests a greater susceptibility to bats due to the presence of many cave shelters that tend to be sensitive to impact with wind farms

(Rodríguez-Durán and Feliciano-Robles, 2015), facts that highlight the high risk of their biodiversity make these areas high risk of impact.

Among the states with parks in the Northeast, Bahia, Rio Grande do Norte and Ceará stand out, with approximately 28, 25 and 15% of the parks in Brazil, respectively. These first two states have intensified their investments in coastal areas and with extensive dune strips, these areas tend to have lower species richness and are therefore less sensitive (Bernard et al., 2014). However, Bahia was considered an area that needs greater concern regarding the presence of wind farms, since it has a large number of projects built and in the process of licensing in various habitats (ANEEL, 2015), in addition to having a deficit of information on the relationship of bats with wind farms (Bernard et al., 2014). In this state, a monitoring carried out in the Sento Sé Region was recorded with high mortality rates, in just one study campaign with two days of samples with an interval of one week, the death of 159 specimens was recorded, with deaths in only 5 of the 43 turbines present in the Wind Farm (personal communication). While in another area of the neotropical region with a warm climate, Puerto Rico, in a two-year monitoring carried out with a fortnightly periodicity in 10 turbines drawn from the 13 present in the park were sampled at each carcass monitoring. This study revealed that 153 bat specimens were found distributed in 11 towers (Rodríguez-Durán and Feliciano-Robles, 2015), confirming the potential risk of these areas.

When evaluating the areas of wind potential in order to analyse the possible expansion scenario of this sector in the country, the southern region of the country also deserves to be highlighted (Amarante et al., 2011), with a forecast of expansion of up to 1.3 megawatts (MW) (EPE, 2015). This area of the country is dominated by species of the Vespertilionidae and Molossidae families that are most affected by collisions with tower blades (Arnett et al., 2016) and has a climate favourable to the migration of these species, which can increase the risk of collision of these individuals (Bernard et al., 2014). In this study, this region presented no
general lower sensitivity values, with the exception of the eastern region of Santa Catarina and southern Paraná.

When evaluating the country's expansion scenario related to the sensitivity found in our results, our analyses showed that only 0.5% of the cells were characterised as conflict areas, that is, there is no conflict between the recorded bat sensitivity and the areas of greatest wind potential in the national territory. Therefore, it is possible to state that there is a possibility of expansion of this sector with reduced impact on the diversity of bats in the country if impact predictability tools such as ours are used, providing guidance for more accurate assessments at local scales, enabling optimal planning of this expansion, since our study indicates areas that need more attention, and consequently rigorous licensing criteria for future projects.

Our results have a clear application for bat conservation in Brazil, through our approach decision makers can use in assessing the susceptibility of areas of the Brazilian territory. The use of our maps as a tool within environmental licensing is able to guide decision makers in environmental management, since their use allows managers to locate potentially high-risk areas for future deployments in this sector, giving indications of locations that should be carefully evaluated at the local scale, in order to assess the real risk generated by the implementation of future projects, thus helping to minimise possible impacts arising from these deployments. Thus, this study aims to contribute to the planning of the construction of new enterprises and in directing preventive measures to reduce the impact of this type of enterprise, serving as a tool for screening the criteria to be taken into account in the licensing of new enterprises. And in guiding the monitoring guidelines of parks already implemented or in the process of licensing.

The environmental licensing of wind farms in the country is governed by the National Environmental Council, CONAMA Resolution No. 462/2014, which establishes procedures for this licensing. Based on this regulation, each state has its environmental agency that will define the Terms of Reference by which each enterprise must follow, depending on its size, which must carry out a Simplified Environmental Report, or Environmental Impact Study, for small parks or for large enterprises respectively.

In this context, our study enables the use of this tool as a way to guide this process, through which areas indicated as potentially conflict areas should be licensed in a more rigorous manner. When the conflict areas are related to the present scenario, parks already installed, this rigour should be related to the monitoring of the enterprise, while when the conflict area is related to the future scenario they should be considered in the licensing process, where the preparation of the Terms of Reference should be more demanding, so that the assessment of the potential impact of this possible enterprise can be verified at the local scale.

Our results also highlighted the value of a cumulative approach to identify areas most important for bats, a crucial piece of information to locate areas sustainably for future wind developments or for other environmental stressors.

We consider our efforts as a first step towards the production of more comprehensive and accurate risk maps, which could be easily implemented using our approach, for example by adding more details related to impacted species in the country and including other locally available information, such as location of large shelters, occurrence of migration corridors, mortality statistics, in order to better understand the characteristics driving bat mortality in wind farms. In this context, our proposal suggests a tool that indicates the potential susceptibility of bats to wind farms. Mapping that has great potential to assist decision makers in analysing methods to be used for the evaluation of environmental impact studies, and may even assist in the creation of guidelines for the environmental licensing of this type of enterprise.

REFERENCES

Brazilian Wind Energy Association. (2013) Our sector. Available at: <http://www.abeeolica.org.br/novo-site/index.php/nosso-setor.html>. Accessed on: 01/12/2015.

Amarante, O.A.C., et al., (2001) Atlas of Brazilian Wind Potential. Available at: <http://www.cresesb.cepel.br/index.php?link=/atlas eolicobrasil/atlas.htm>. Accessed on: 08/11/2015.

National Electric Energy Agency (2015) Eolioelectric Generating Centres - EOL. Georeferenced Information System of the Electricity Sector - SIGEL. Available: <http:// sigel.aneel.gov.br/sigel.html>. Accessed on: 15/09/2016.

Arnett, E.B. & Baerwald, E.F. (2013) Impacts of Wind Energy development on Bats: Implications for Conservation. In: Rick A. A.; Scott C. P..Bat Evolution, Ecology, and Conservation. *Springer Open*. New York, 435-456.

Arnett,, E.B., Baerwald, E.F., Mathews, F., Rodrigue, L., Rodríguez-Durán, A., Rydell, J., Villegas-Patraca, R., Voigt, C.. (2016). Impacts of Wind Energy development on bats: A global perspective. In: Voig, C.C., Kingston, T.. Bats in the Anthropocene: Conservation of bats in a changing the world. Springer Open.

(2015) Wind Turbine Interactions with Wildlife and their Habitats: A summary of Research Results and priority questions. Available at: < https://awwi https://awwi.org/wp-content/uploads/2016/07/AWWI-Wind-Wildlife-Interactions- Summary-June-2016.pdf>. Accessed on: 16/11/2015.

Bader, E., Jung K., Kalko, E. K. V., Page, R., Rodriguez, R., Sattler, T. (2015) Mobility explains the response of aerial insectivorous bats to anthropogenic habitat change in the Neotropics. *Biological conservation*. 186:97-106.

Baerwald, E.F., & Barclay, R.M.R. (2009) Geographic variation in activity and fatality of migratory bats at wind-energy facilities: Journal of Mammalogy, v. 90, p. 1341-1349.

Barros, M. A. S., R. Gastal De Magalhaes, and A. M. Rui (2015) Species composition and mortality of bats at the Osorio Wind Farm, southern Brazil. Stud. on Neotr. Faun. and Envir., 50: 31-39.

Bernard E. (2001) Vertical stratification of bat communities in primary forests of Central Amazon, Brazil. J. Trop. Ecol. 17: 115-126.

Bernard, E. (2002). Diet, activity and reproduction of bat species (Mammalia, Chiroptera) in Central Amazonia, Brazil. Ver. Bras. de Zool., 19(1), 173-188.

Bernard, E., Machado, R.B. & Aguiar, L.M. (2011) Discovering the Brazilian Bat fauna: a task for two centuries? Mammal Rev. 41 (1):23-39.

Bernard, E., et al. (2012) A horizon scan on bat conservation in Brazil. In: Freitas, T.R.O., Vieira, E.M. (Eds.), Mammals of Brazil: Genetics, Systematics. Ecology and Conservation, vol. II. Soc. Bras. Mastoz., Rio de Janeiro. 19-35.

Bernard, E., Paese, A., Machado, R. B., Aguiar, L.M.S. (2014). Blown in the Wind:bats and Wind farms in Brazil. Nat. and cons. (12):106-111.

Bevanger K. (1994) Bird interactions with utility structures: collision and electrocution, causes and mitigating measures. Ibis, **136,** 412-425.

Boitani, L., Maiorano, L., Baisero, D., Falcucci, A.,Visconti, P. & Rondinini, C. (2011) What spatial data do we need to develop global mammal conservation strategies? Phil. Trans. R. Soc. B (366):2623-2632.

Bradbury G, Trinder M, Furness B, Banks AN, Caldow RWG (2014) Mapping Seabird Sensitivity to Offshore Wind Farms. PLoS ONE 9(9).

Brazil. Ministry of the Environment, National Environmental Council, CONAMA. CONAMA Resolution n°462/2014, of 24 July 2014 - In: Resolution, 2014.

Available at:< http://www.mma.gov.br/port/conama/legiabre.cfm?codlegi=703cfm?codlegi=703> Accessed on: 16/01/2017.

Brazil. Law No. 12,651, 25 May 2012. Provides for the New Forest Code. Accessed on: 05/11/2016.

Buckley LB, Davies T.J, Ackerly Dd, Kraft NJB, Harrison SP, Anacker BL, Cornell HV, Damschen EI, Grytnes JA, Hawkins BA, Mccain CM,Stephens PR & Wiens JJ. (2010) Phylogeny, niche conservatism and the latitudinal diversity gradient in mammals. Proceedings of the Royal Society B: Biol. Scienc..277: 2131-2138.

Calenge, C. (2006) The package adehabitat for the R software: a tool for the analysis of space and habitat use by animals. Ecol. Mod., 197, 516-519.

Castelletti C.H.M., Silva JMC, Tabarelli M, Santos A. (2004) How much is left of the Caatinga? A preliminary estimate. In: Silva JMC, Tabarelli M, Lins L (eds) Biodiversity of the Caatinga: priority areas and actions for conservation. Ministry of the Environment, Brasilia, pp 92-10.

Cianciaruso, M. V., Silva, I. A., & Batalha, M. A. (2009). Phylogenetic and functional diversity: new approaches for community ecology. Biot. Neot.,9(3), 93-103.

Clifford, P., Richardson, S. & Hemon, D. (1989) Assessing the significance of the correlation between 2 spatial processes. Biomet., 45, 123-134.

Geological Survey of Brazil: Companhia de Pesquisa de Recursos Minerais (CPRM). Available at: http://www.cprm.gov.br/publique/cgi/cgilua.exe/sys/start.htm?infoid=1481&sid=9/cgilua.exe/sys/start.htm?infoid=1481&sid=9(Accessed on: 28/06/2016).

Daniel J. Stekhoven .(2013) MissForest: Nonparametric Missing Value Imputation using Random Forest.R package version 1.4.

Denzinger A. & Schnitzler H. (2013). Bat guilds, a concept to classify the highly diverse foraging and echolocation behaviours of microchiropteran bats. Front. in Phys. v 4.

Di Gregorio, A.; Jansen, L. J. M. (2000) Land cover classification system: classification concepts and user manual. Available at: http://www.fao.org/DOCREP/003/X0596E/X0596e00.htm (accessed: 23/11/2015).

Empresa de Pesquisa Energética. (2014) Energy demand studies 2050. Rio de Janeiro/RJ.

Empresa de Pesquisa Energética. (2015) Ten-year Energy Expansion Plan 2024. Brasília. Accessed on   : 15.10.2016.   Available at   : http://www.epe.gov.br/PDEE/Relat%C3%B3rio%20Final%20do%20PDE%202024.pdf.

European Space Agency. (2010) Global Land Cover Map version .2.3. Accessed on: 18.08.2015. Available at: http://due.esrin.esa.int/page globcover.php.

Farneda, F. Z., R. Rocha, A. López -Baucells, M. Groenenberg, I. Silva, J. M. Palmeirim, P. E. D. Bobrowiec, and C. F. J. Meyer (2015). Trait -related responses to habitat fragmentation in Amazonian bats. Joun. of Appl. Ecol.

Feijó A, Rocha Pa, Althoff Sl. (2015) New species of Histiotus (Chiroptera: Vespertilionidae) from northeastern Brazil. Zoot. 4048:412-427.

Ferrara Fj, Leberg Pl. (2005) Characteristics of positions selected by day-roosting bats under bridges in Louisiana. Journ. of mam..86:729-735.

Ferreira, D., Freixo, C., Cabral, J.A., Santos, R., Santos, M., (2015) Do habitat characteristics determine mortality risk for bats at wind farms? Modelling usceptible species activity patterns and anticipating possible mortality events. Eco. Inform. 28, 7-18.

GAO. (2005) US Government Accountability Office, Windpower: Impacts on wildlife and government responsibilities for regulating development and protecting wildlife.Washington, DC: US Government Accountability Office,Report GAO-05-906.

Hull Cl; Cawthen L. (2013) Bat fatalities at two wind farms inTasmania, Australia: bat characteristics, and spatial and temporal patterns. N Z J Zool 40(1):5-15. doi:10.1080/03014223.2012.731006

International Agency Energy. World energy outlook 2014 factsheet: Power and renewables. Available from: http://www.worldenergyoutlook.org/media/wewebsite/2014/141112 WEO FactSheets. pdf. Accessed on: 10/12/2015.

Instituto Chico Mendes de Conservação da Biodiversidade. (2016) Annual report of routes and areas of concentration of migratory birds in Brazil. Cabedelo: PB: CEMAVE/ ICMBio. ISSN: 2446-9750. Accessed on: 15/11/2016. Available at: http://www.icmbio.gov.br/portal/images/stories/DCOM_Miolo_Rotas_Migrat%C3%B3_rias_2016 final.pdf.

Brazilian Institute of Geography and Statistics. 2007. Manual técnico de pedologia, Coordenação de Recursos Naturais e Estudos Ambientais. 2. Ed. - Rio de Janeiro: IBGE. (Manuais técnicos em geociências, ISSN 0103-9598; n4).

International Union For Conservation Of Nature. 2016. The IUCN Red List of Threatened Species. Version 2014.1. Available at: http://www.iucnredlist.org. Accessed on: 16/11/2015.

IUCN Standards and Petitions Subcommittee (2016). Guidelines for Using the IUCN Red List Categories and Criteria. Version 12. Prepared by the Standards and Petitions Subcommittee. Downloadable from http://www.iucnredlist.org/documents/RedListGuidelines.pdf.

Jacobson, M.Z. (2009). Review of solutions to global warming, air pollution, and energy security. Ener. and Environm. Scienc, 2, 148-173.

Jenkins A.R., Smallie J.J., & Diamond M. (2010) Avian collisions with power lines: a global review of causes and mitigation with a South African perspective. Bird Conserv. Inter., 20, 263-278.

Jetz W, Freckleton Rp (2015) Towards a general framework for predicting threat status of data-deficient species from phylogenetic, spatial and environmental information. Philosophical Transactions of the Royal Society B: Biol.l Scienc. 370: 1-10.

Johnson GD, Perlik MK, Erickson WP, Strickland MD. (2004) Bat activity, composition, and collision mortality at a large wind plant in Minnesota. Wildlife Soc B. 32:1278-1288.

Joint Nature Conservation Committee (JNCC). Accessed on: 05/12/2015. Available at: http://jncc.defra.gov.uk/page-2273.

Jones, K.E, Purvis A, Gittleman J.L. (2003) Biological correlates of extinction risk in bats. Am. Nat. 161, 601-614 93 Duchamp, J.E. and Swih

Jung, K., & Kalko, E. K. (2011). Adaptability and vulnerability of high flying Neotropical aerial insectivorous bats to urbanisation. Diversity and Distributions, 17(2), 262-274.

Kalka, MB, Smith AR, Kalko EKV (2008) Bats limit arthropods and herbivory in a tropical forest. Scienc. 320: 71-71.

Kalko, E. K. V., Handley C. O., Handley D. (1996). Organisation, diversity, and long-term dynamics of a neotropical bat community, in Long-term Studies in Vertebrate Communities, eds Cody M., Smallwood J., editors. Los Angeles, CA: Acad. Press. 503553

Kerth, G, Almasi B, Ribi N, Thiel D, L"Upold S. (2003). Social interactions among wild female Bechstein's bats (Myotis bechsteinii) living in a maternity colony. Acta Ethol. 5:107- 114.

Klingbeil, B. T. & Willig, M.R. (2009). Guild-specific responses of bats to landscape composition and configuration in fragmented Amazonian rainforest. J App Ecol 46:203-213.

Kunz, Thomas, Arnett, E.B., Erickson,W.P., Hoar,A.R., Johnson, G.D., Larkin, R.P., Strckland, M.D., Thresher, R., Tuttle, M.D. (2007) Ecological impacts of wind energy development on bats: questions, research needs and hypotheses. Front. Ecol. Environ; 5 (6): 315-324.

Lemes P. Faleiro F.A.M.V., Tessarolo G. & Layola R.D. (2011) Refining Spatial Data for Biodiversity Conservation. Nat. and Conserv., 9(2):240-243.

McGuinness, S., Muldoon C, Tierney N, Cummins S, Murray A, Egan S, Crowe O (2015) Bird sensitivity mapping for wind energy developments and associated infrastructure in the Republic of Ireland. BirdWatch Ireland, Kilcoole, Wicklow

Milner, J., C. Jones, and J. K. Jones, Jr. (1990). *Nyctinomops macrotis*. Am. Soc. Mamm., Mammalian Species No. 351:1-4.

Moratelli, R, Dias D. (2015) A new species of nectar-feeding bat, genus Lonchophylla, from the Caatinga of Brazil (Chiroptera, Phyllostomidae). ZooKeys. 514:73-91.

Myers, N; Mittermeier, R. A; Mittermeier, C. G; Fonseca, G. A. B; Kent, J. (2000). Biodiversity hotspots for conservation priorities. Nat., n 403, p.853-859.

National Research Council. (2007) Environmental Impacts of Wind-Energy Projects, Washington, DC: The Natio. Acad. Press.

Neuweiler, G.. (2000) The biology of bats. New York: Oxford University Press, 310p

Nogueira, M.R., Lima, I. P., Moratelli, R., Tavares, V. C., Gregorin, R., Perachi, A. L. (2014) Check list o f Brazilian bats, with comments on original records. Chec. list. v. 10, n° 4: p. 808-821.

Norberg, U. M. & Rayner, J. M. V. (1987) Ecological morphology and flight in bats (Mammalia; Chiroptera): wing adaptations, flight performance, foraging strategy and echolocation. Phil. Trans. R. Soc. (B) 316:

335-427.

Norberg, U.M..(1994) Wing design, flight performance, and habitat use in bats. Ecological morphology: integ. organ. biol. Pp.205-239.

(2012) Guidelines for Interpreting Listing Criteria for Species, Populations and Ecological Communities under the NSW Threatened Species Conservation Act, Version 1.3. January 2012. Office of Environment and Heritage, Sydney NSW

Okey, B. W., & Kuzemchak, M. L. (2009) Modelling potential wildlife-wind energy conflict areas. Harrisburg: The Center for Rural Pennsylvania. Accessed on: 10/10/2016. Available at: http://www.rural.palegislature.us/wind wildlife report09.pdf

Oliveira-Galvão A.C (2014) The geospatial database of the national centre for cave research and conservation. RBEsp.v.1,n°4.

Pacheco S.M.; Marques R.V.; Esbérard C.E. (2008) Bats of Brazil: Biology, Ecology and Conservation. Armazém Digital, Porto Alegre, Brazil. 235-244.

Pacheco, S.M., Barros, M., Azevedo, V.L. Y Etges, M. (2014) Wind farms and the conservation of bats in Brazil. Paper presented at the symposium "Wind energy development and its impact on the bats of Latin America and the Caribbean: foundations for the establishment of impact assessment guidelines in RELCOM".I COLAM, Quito, Ecuador.

Peracchi, A.L.; Nogueira, M.R. (2007) Order Chiroptera. In: Reis, N. R.; Peracchi, A. L; Pedro, W. A.; Lima, I. P.. (Eds.). Bats of Brazil. 1 ed. Londrina. 27-36.

Penone, C, Davidson Ad, Shoemaker Kt, Di Marco M, Rondinini C, Brooks Tm, Young Be, Graham Ch, Costa Gc. (2014) Imputation of missing data in lifehistory traits datasets: which approach performs the best? Methods in Ecology and Evolution, 5, 961-970.

Piorkowski MD, O'Connell TJ. (2010) Spatial pattern of summer bat mortality from collisions with wind turbines in mixed-grass prairie. Am Midl Nat. 164:260-269.

Quinn, M., Alexander, S., Heck, N., & Chernoff, G. (2011). Identification of bird collision hotspots along transmission power lines in Alberta: an expert-based geographic information system (GIS) approach. Journ. of Environm. Infor, 18(1).

Rangel, T.F.L.V.B., Diniz-Filho, J.A.F., Bini, L.M. (2010) SAM: a comprehensive application for spatial analysis in macroecology. Ecography 33, 46-50. Available at: www.ecoevol.ufg.br/sam.

Robert J. Hijmans. (2016). raster: Geographic Data Analysis and Modelling .R package version 2.5-8. https://CRAN.R-project.org/package=raster

Rodríguez-Dúran, A. (2009). Bat assemblages in the West Indies: the role of caves. Pp. 265-280, in: Fleming T. H.; Racey P. A.. Island bats: Evolution, Ecol. and Conserve. University of Chicago Press, Chicago, 560.

Rodriguez-Dúran, A; Feliciano-Robles, W. (2016). Impact of Wind Facilities on Bats in the Neotropics. Acta Chiropt., 17(2): 365-370.

Rondinini, C., Di Marco, M., Chiozza. (2011). Global habitat suitability models of terrestrial mammals. Roy. Soc. Biol. Scienc.

Roscioni, F., Russo, D., Di Febbraro, M., Frate, L., Carranza, M. L.; Loy, A. (2013) Regional- scale modelling of the cumulative impact of wind farms on bats. Biod. and Conserv.22(8), 1821-1835.

Roscioni F, Rebelo H, Russo D, Carranza ML Di Febbraro M, Loy A (2014) A modelling approach to infer the effects of Wind farms on landscape connectivity for bats. Landscape Ecol.

Rydell, J, Bach L, Dubourg-Savage M, Green M, Rodrigues L, Hedenstrom A. (2010) Bat mortality at wind turbines in northwestern europe. Acta Chiropterolog 12(2):261-274.

Safi, K., & Kerth, G. (2004). A comparative analysis of specialisation and extinction risk in temperate- zone bats. Conserv. Biol., 18(5), 1293-1303.

Sagot, M., & Chaverri, G. (2015). Effects of roost specialisation on extinction risk in bats. Conserv. Biol. 29(6), 1666-1673.

Santos H, Rodrigues L, Jones G, Rebelo H (2013) Using species distribution modelling to predict bat fatality risk at wind farms. Biol Conserv 157:178-186.

Schuster, E., L. Bulling, and J. Koppel. (2015). Consolidating the state of knowledge: a synoptic review of wind energy's wildlife effects. Env. Manag. 56:300-331.

Silva, J. M. C., M. Tabarelli, M. T. Fonseca, and L. Lins. 2004. Biodiversity of the Caatinga: priority areas and actions for conservation. Ministry of the Environment, Brasília.

Simmons, N. B. (2005). Order chiroptera. Mammal species of the world: a taxonomic and geographic reference,1, 312-529.

Sovernigo, M. H. (2009). Impact of Wind Turbines on Avifauna and Chiropterofauna in Brazil. Dissertation (Course Conclusion Work) - Federal University of Santa Catarina - Department of Ecology and Zoology. Santa Catarina.

Stevens, Rd. (2004) Untangling latitudinal richness gradients at higher taxonomic levels: familial perspectives on the diversity of New World bat communities. Journ.of Biogeog. 31: 665674.

Voig, C.C., Kingston, T.. (2016) Bats in the Anthropocene:Conservation of bats in a changing the world. Spring. Open.

Wilman, H., Baelmaker, J., Simpson, C. R., Rivedeneira, M.M., Jetz, W. . (2014) Eltontraits 1.0: Species-level foraging attributes of the world's birds and mammals: Ecological archives e095-178. Ecol. 95: 2027-2027.

World Wildlife Fund for Nature (WWF). Available at: http://www.wwf.org.br/?40562/more- clean-energy-and-less-environmental-impact (Accessed: 10/12/2015).

I want morebooks!

Buy your books fast and straightforward online - at one of world's fastest growing online book stores! Environmentally sound due to Print-on-Demand technologies.

Buy your books online at
www.morebooks.shop

Kaufen Sie Ihre Bücher schnell und unkompliziert online – auf einer der am schnellsten wachsenden Buchhandelsplattformen weltweit! Dank Print-On-Demand umwelt- und ressourcenschonend produziert.

Bücher schneller online kaufen
www.morebooks.shop

FSC
www.fsc.org
MIX
Papier aus verantwortungsvollen Quellen
Paper from responsible sources
FSC® C105338